THE WONDER OF UNDERWATER ANIMALS

THE WONDER OF UNDERWATER ANIMALS

Published by Fog City Press,
a division of Weldon Owen Inc.
1045 Sansome Street
San Francisco, CA 94111 USA

www.weldonowen.com

weldon**owen**

President & Publisher Roger Shaw
Associate Publisher Mariah Bear
SVP, Sales & Marketing Amy Kaneko
Finance Manager Philip Paulick
Editor Bridget Fitzgerald
Creative Director Kelly Booth
Art Director Meghan Hildebrand
Senior Production Designer Rachel Lopez Metzger
Production Director Chris Hemesath
Associate Production Director Michelle Duggan
Director of Enterprise Systems Shawn Macey
Imaging Manager Don Hill

Weldon Owen is a division of **BONNIER**

Library of Congress Control Number on file with the publisher.

ISBN 13: 978-1-68188-094-5
ISBN 10: 1-68188-094-6

10 9 8 7 6 5 4 3 2 1

2016 2017 2018 2019

Printed by 1010 Printing in China.

From shallow coral reefs to deep murky trenches, the ocean offers an underwater realm that's home to many remarkable animals and plants.

The ocean boasts a vast array of species, from forests of plankton (made up of tiny plants that produce half the oxygen we breathe), to massive creatures that dominate the seas.

The ocean covers nearly three-fourths of our planet's surface. Maybe we should really call it Planet Ocean!

Underwater animals come in all shapes and sizes—even ones that don't look like animals at all.

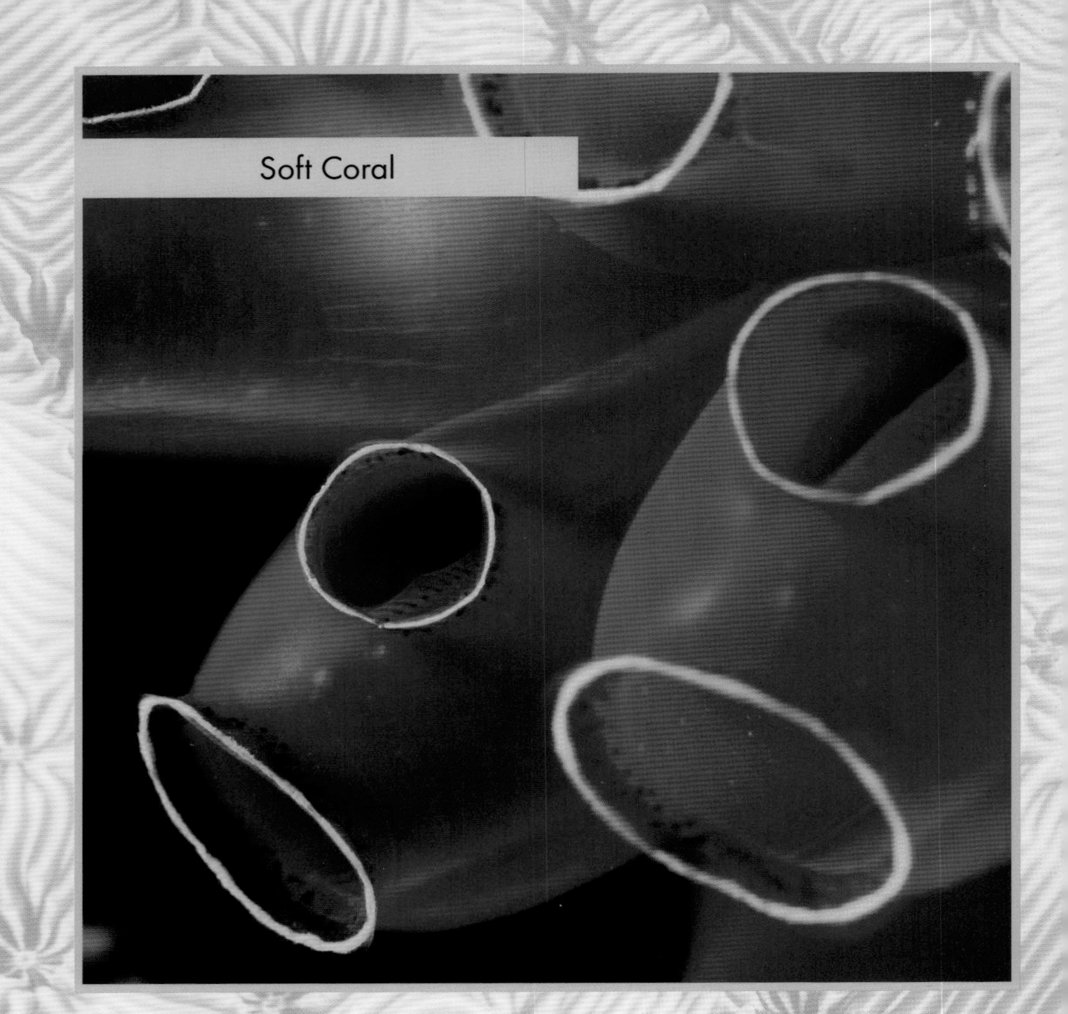
Soft Coral

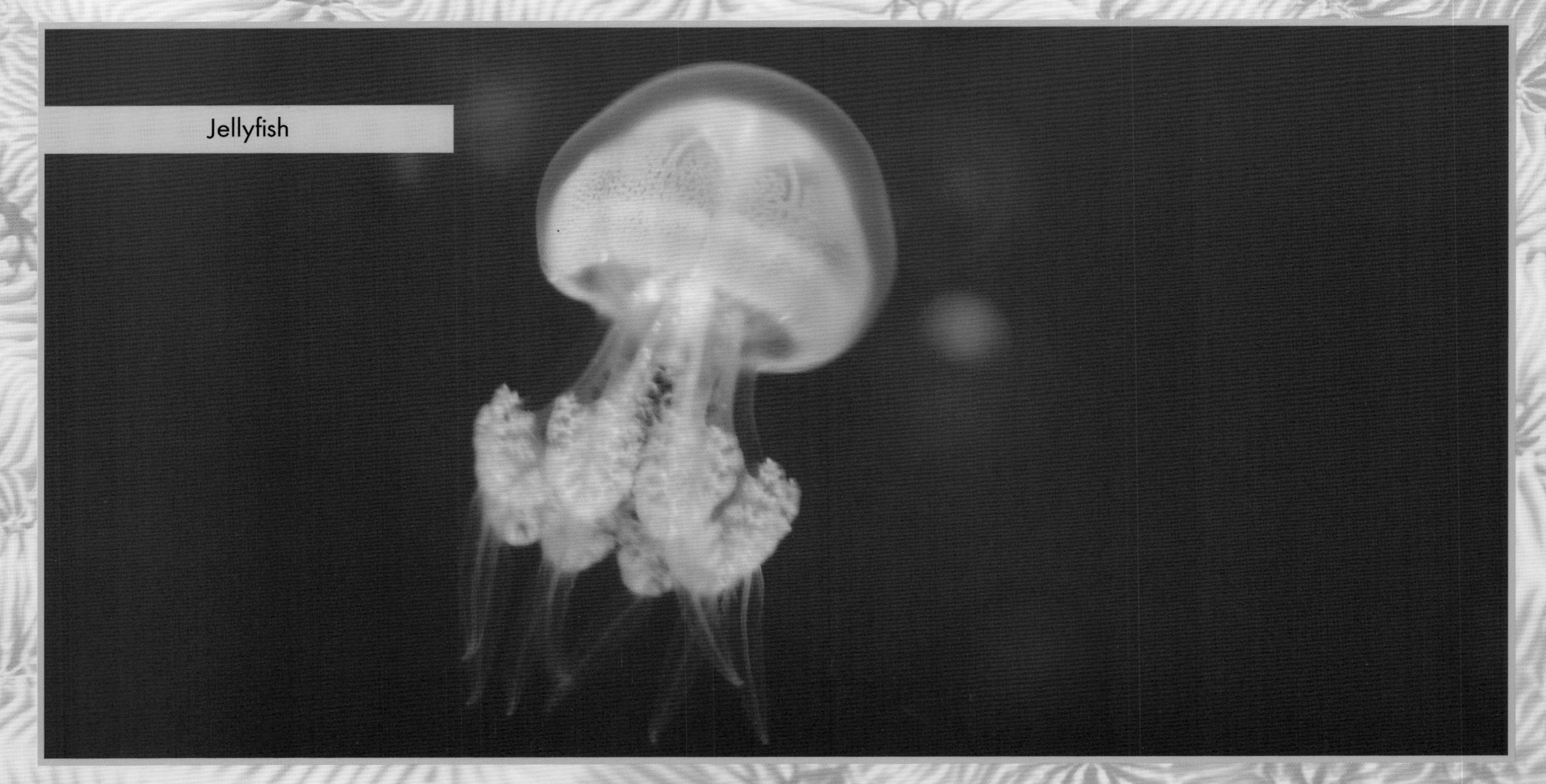
Jellyfish

Tube Anemone

Fun Fact

Orcas are fierce whales that can even eat sharks.

Whale Shark

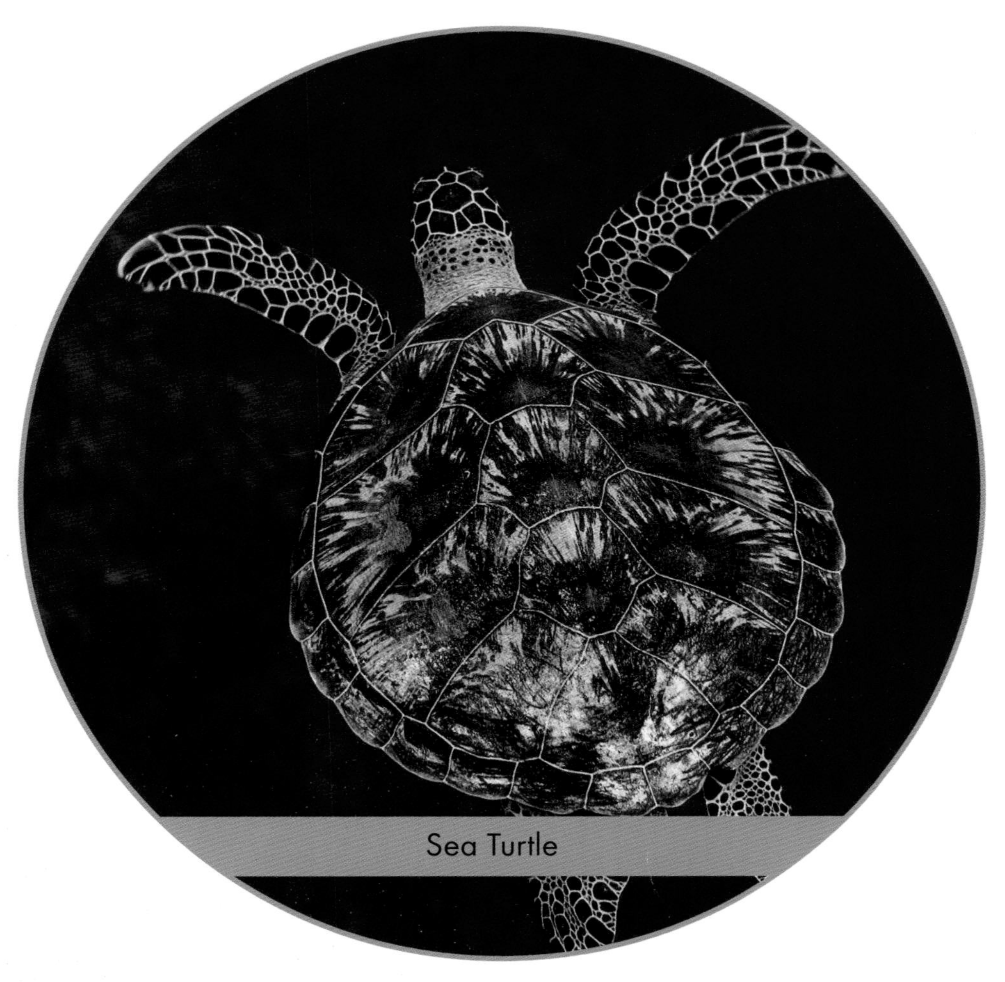
Sea Turtle

Orcas, sea turtles, and whale sharks are some of the creatures that make their homes in the ocean.

Sea star

Caribbean Octopus

Some animals, such as sea stars and octopuses, have long arms or tentacles to help them grip the sea floor—or their prey.

Feather Duster

These "feather dusters" are really a type of worm. They use their fans to gather food and oxygen.

Tube Anemone

Feather Duster

Red Slate Pencil Urchin

Fun Fact

These spiny urchins can be as long as your arm!

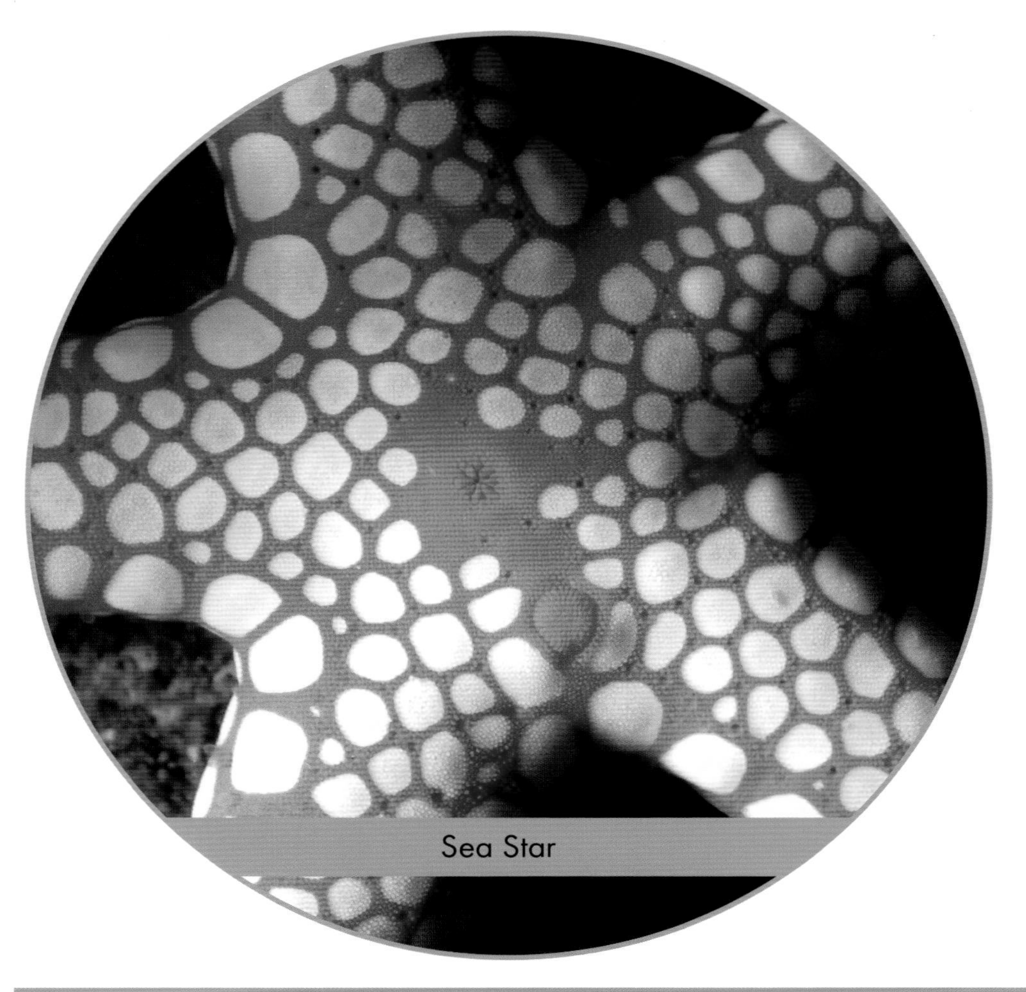
Sea Star

Some sea life come in stunning colors. Most live on the continental shelf, the part of the ocean closest to land.

Nudibranch

Red Frogfish

Camouflaging colors help animals hide from predators. Sometimes it also helps them hide from their prey—until it's too late!

Decorator Crab

Fun Fact

Grown sea turtles have very few predators.

Sea Turtle

Spotted Cleaner Shrimp

Many ocean creatures look like no other animal on Earth, such as this spotted cleaner shrimp!

Fun Fact

Some fish and shrimp live inside sea anemones.

Spotted Cleaner Shrimp

Fun Fact

Jellyfish come in many shapes, colors, and sizes.

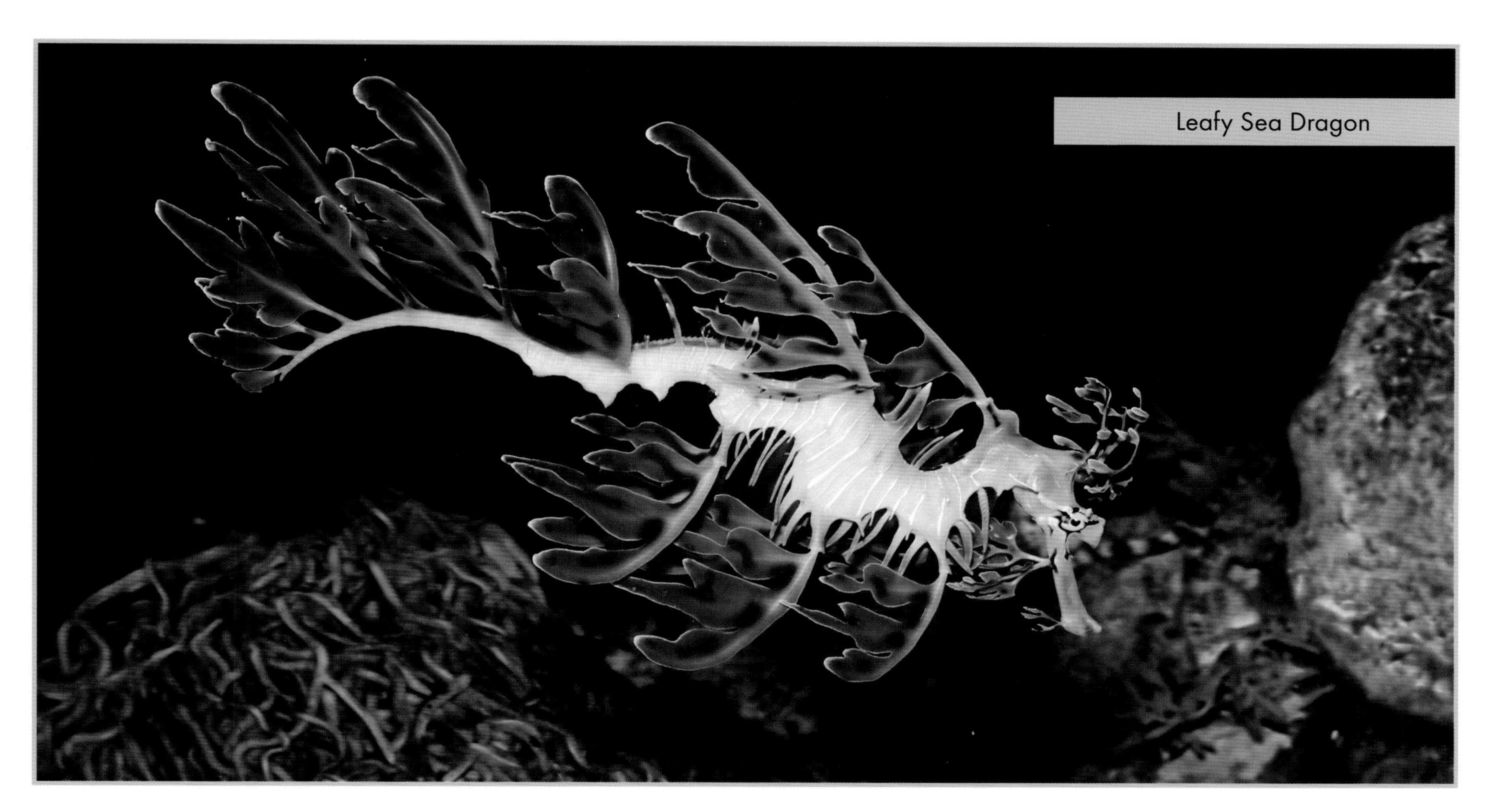
Leafy Sea Dragon

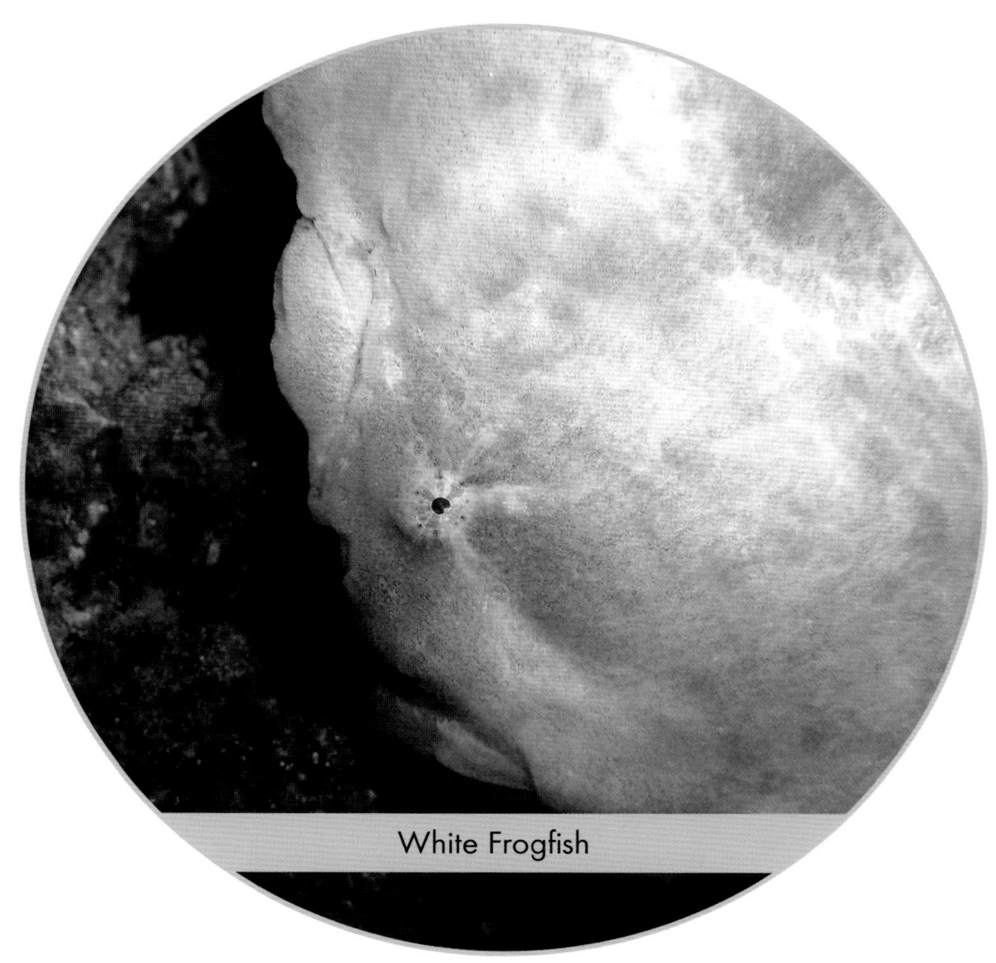
White Frogfish

Other underwater animals look exactly like surrounding rocks, leafy plants . . . or even spaceships!

Sea creatures are not always friendly—they must know how to fight in order to survive.

Giant Octopus

Spider Crab

Coconut Octopus

White-Spotted Octopus

Giant Pacific Octopus

Octopuses squirt ink at their enemies, allowing the octopus to hide. They have also been known to fight small sharks—and win!

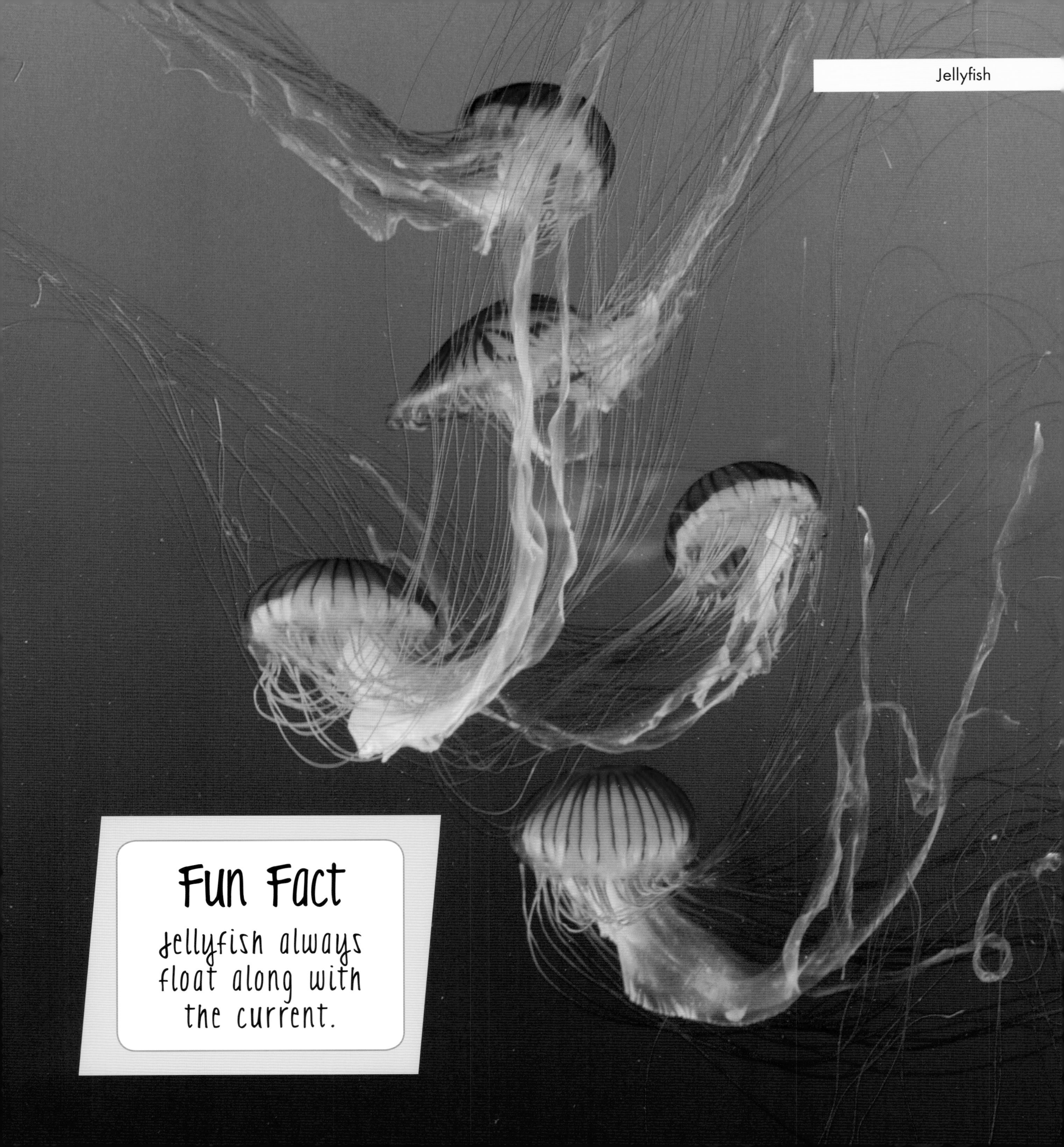

Fun Fact

Jellyfish always float along with the current.

Moon Jellyfish

Pacific Moon Jellyfish

Some jellies may be beautiful, but their tentacles often contain poison that will sting if it touches your skin.

Hermit Crab

Crabs, which may seem small and harmless under their shells, are armed with pinching claws.

Horseshoe Crab

Fun Fact

Horseshoe crabs are most closely related to spiders!

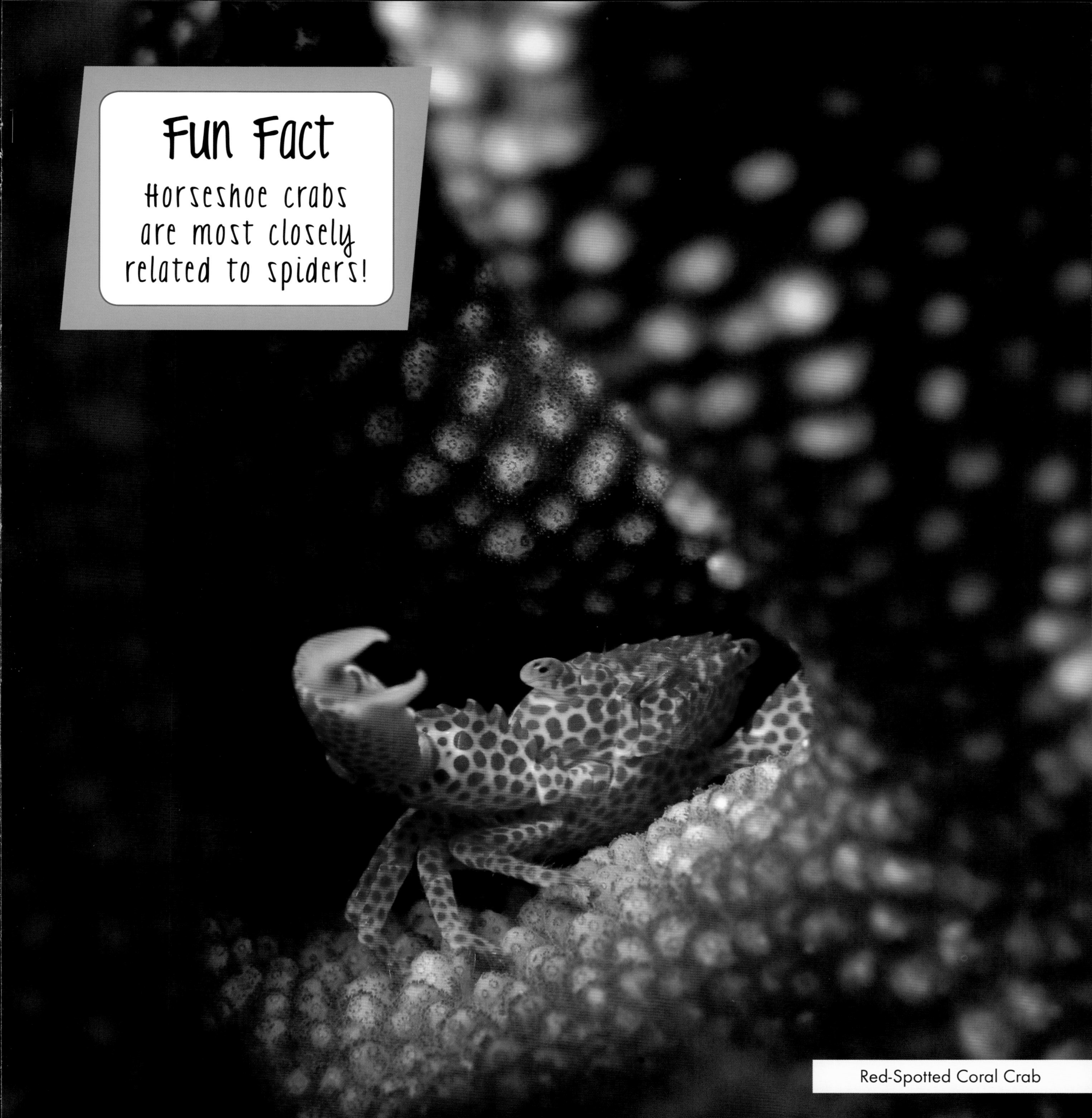

Red-Spotted Coral Crab

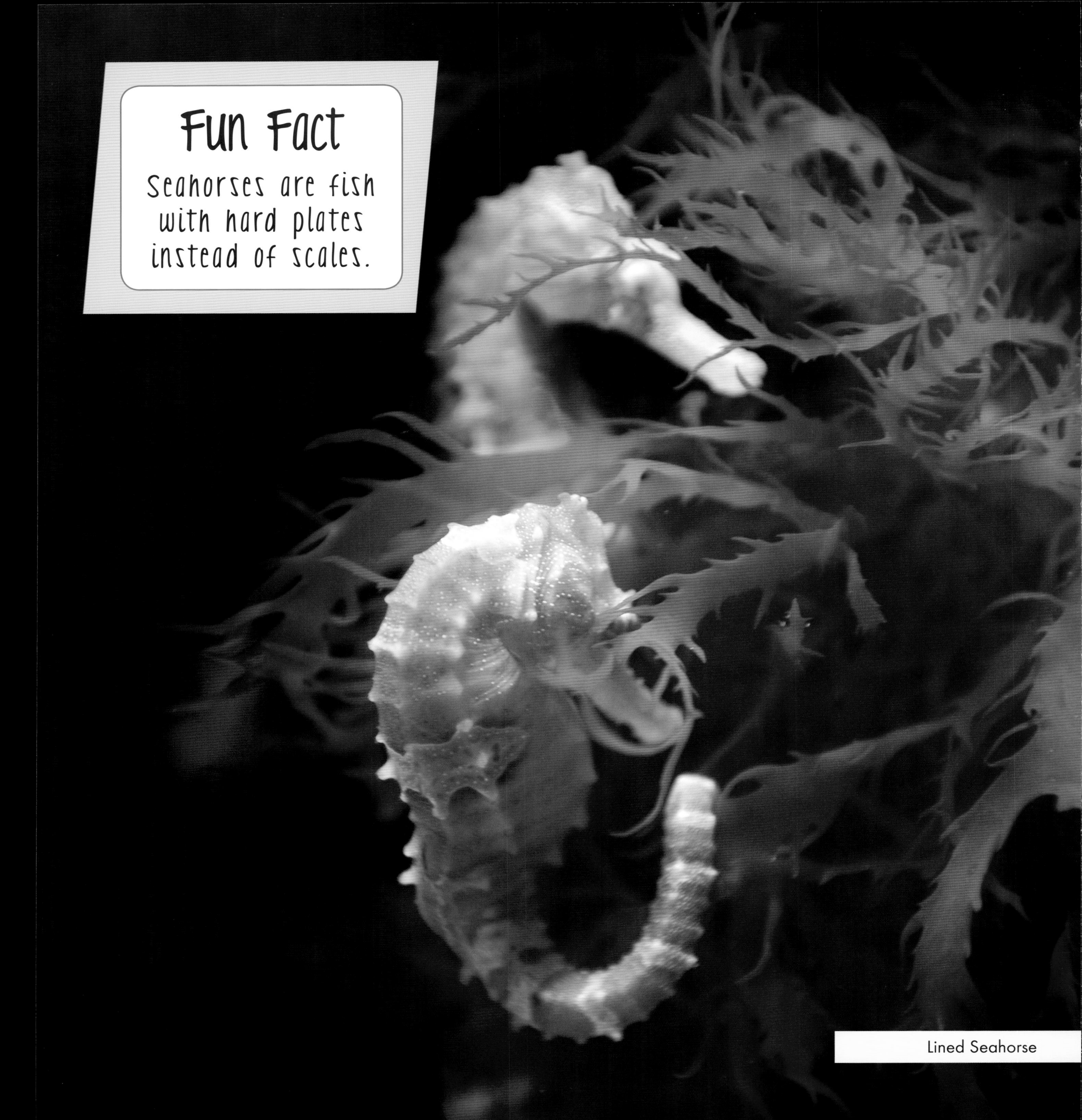

Fun Fact

Seahorses are fish with hard plates instead of scales.

Lined Seahorse

Moray Eel

With so many hidden dangers, the ocean may seem like a hazardous environment.

Great White Shark

Coral

And yet, the ocean can be such a beautiful and peaceful place.

Sea Turtle

Whale Shark

Some of the ocean's largest animals aren't very dangerous at all. Whale sharks eat only microscopic plankton and some small fish.

Whale Shark

Whale Shark

Fun Fact

Whales are mammals that surface to breathe.

Gray Whale

Beluga Whale

Most whales eat plankton. It's amazing that such large animals can survive on a diet of tiny creatures.

Minke Whale

Fun Fact

Red rock urchins are also called rock-boring urchins.

Red Rock Urchin

Long=Spined Sea Urchin

This animal may look like a pincushion, but it's actually a sea urchin.

Just like us, ocean animals can often be found hanging out together, with families and friends.

Seahorse

Bottlenose Dolphin

Fun Fact

Blue blubbers can also be white or purple.

Blue Blubber Jellyfish

Horseshoe Crab

The ocean is the one habitat on Earth where humans do not live. Perhaps this is what makes it so fascinating to us.

Fun Fact

There are about 2,000 species of sea stars!

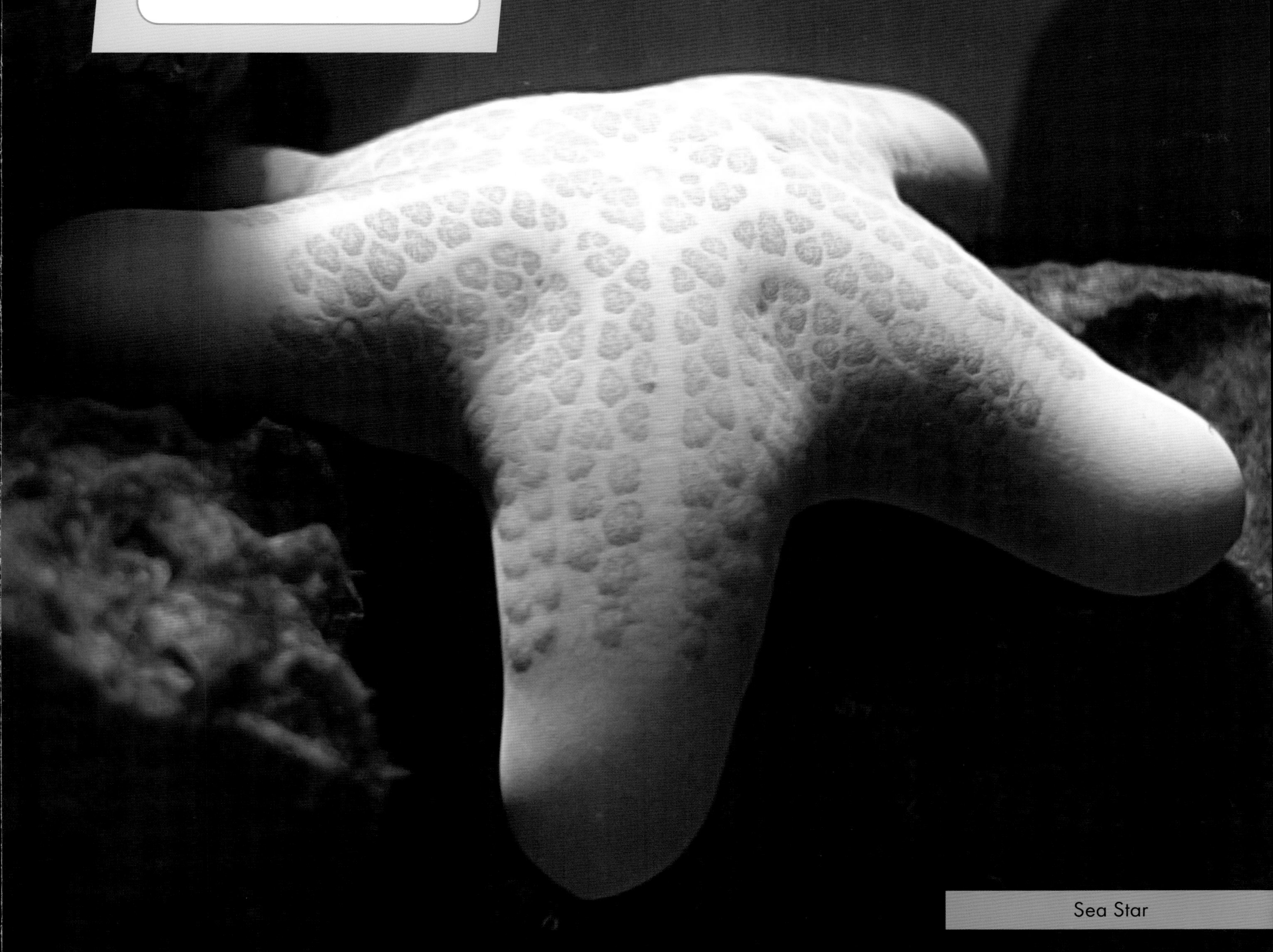

Sea Star

Sea Turtle

Blue-Ridged Octopus

Blue-Spotted Stingray

Luckily, we can visit and observe some of the amazing animals that live here. What a mysterious place to explore!

Mantis Shrimp

Key: SS=Shutterstock; iSP=iStockphoto; DT=Dreamstime;
t=top; b=bottom

SS: 2, 5, 6, 9, 10, 11t, 12, 16, 18t, 19, 20, 26, 28, 30t, 33b, 34, 35, 36t, 36b, 37, 38, 39t, 40, 41, 42t, 42b, 43, 44, 45, 46, 48.
iSP: 8t, 8b, 11b, 13, 14t, 14b, 15, 17b, 24t, 25, 27, 29t, 29b, 33t, 39b.
DT:17t, 18b, 21, 22, 23t, 23b, 24b, 30b, 31, 32, 47t, 47b.